Table of Content

Introduction

Well control is one of the important topics which several positions in the drilling industry must fully understand. The well control quiz ebook is created to help people refresh and get more understanding about well control. It will also assist people before they are going to the well control school. There are several books in this series and each book contains one main topic which will be advantageous for learners.

The eBook is an interactive eBook containing 25 questions related to **basic pressure in well control** covered from the basic to some harder calculations. Additionally, this content is based on both IWCF and IADC well control. Each question contains 3 possible answers. If learners select the right one, they will allow going to the next question. What's more, each question will have the full explanations. As a minimum, you will need a pen or pencil, paper, well control formula and calculator.

Let's get started.

Question - 1: Well information: TD at 8,000'MD/7,500' TVD. Current mud weight is 9.8 ppg. What is the hydrostatic pressure?

a) 3,822 psi

b) 4,077 psi

c) 5,200 psi

This answer is incorrect.

Please try again.

Go back to the question#1

This is the correct answer.

Hydrostatic pressure = 0.052 x 7500 x 9.8 =3,822 psi

Go to next question.

Question - 2: What is Equivalent Mud Weight in ppg for a formation with 0.62 psi/ft?

a) 11.9 ppg

b) 12.9 ppg

c) 13.2 ppg

This answer is incorrect.

Please try again.

Go back to the question#2

☻ **This is the correct answer.**

Equivalent Mud Weight in ppg = 0.62÷0.052 =11.9 ppg

Go to next question.

Question - 3: What is the pressure gradient (psi/ft) for 13.0 ppg mud?

a) 0.676 psi/ft

b) 0.732 psi/ft

c) 0.841 psi/ft

This answer is incorrect.

Please try again.

Go back to the question#3

This is the correct answer.

Pressure gradient (psi/ft) = 13.0 x 0.052 = 0.676 psi/ft

Go to next question.

Question - 4: Hole depth = 9,000'MD/8,500'TVD.
Current mud weight is 9.2 ppg.
Annular pressure loss = 400 psi.
What is the ECD at the bottom of the well?

a) 10.1 ppg
b) 11.1 ppg
c) 12.1 ppg

This answer is incorrect.

Please try again.

<u>Go back to the question#4</u>

This is the correct answer.

ECD = Current Mud Weight + (Annular pressure loss ÷ 0.052 ÷ TVD)

ECD = 9.2 + (400÷0.052÷8500) = 10.1 ppg

Go to next question.

Question - 5: Mud gradient = 0.83 psi/ft. What is the bottom hole pressure at 9,800' MD/9,000' TVD?

a) 8,134 psi

b) 7,470 psi

c) 7,221psi

This answer is incorrect.

Please try again.

<u>Go back to the question#5</u>

This is the correct answer.

Hydrostatic pressure at the bottom hole = 0.83 psi/ft x 9,000' TVD = 7,470 psi.

Go to next question.

Question - 6: Formation pressure is 5,500 psi and the formation is at 8,000'MD/7,500' TVD. What is the pressure in term of mud weight (ppg)?

a) 13.2 ppg
b) 13.5 ppg
c) 14.1 ppg

This answer is incorrect.

Please try again.

Go back to the question#6

This is the correct answer.

Mud weight (ppg) = 5,500 ÷0.052÷7,500 = 14.1 ppg

<u>Go to next question.</u>

Question - 7: Formation top = 12,000'MD / 11,000 TVD.
Reservoir pressure is 6,000 psi.
What is minimum mud weight for drilling without allowing influx into the wellbore?

a) 10.5 ppg

b) 10.0 ppg

c) 9.6 ppg

This answer is incorrect.

Please try again.

Go back to the question#7

 **This is the correct answer.**

Mud weight should be at least balance formation pressure.
Mud weight = 6,000÷0.052÷11,000 = 10.5 ppg.

<u>**Go to next question.**</u>

Question - 8: What are factors contributing to bottom hole pressure when the well is shut in?

a) Surface pressure and annular pressure loss
b) Hydrostatic pressure and annular pressure loss
c) Hydrostatic pressure and surface pressure

This answer is incorrect.

Please try again.

Go back to the question#8

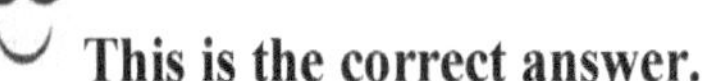 **This is the correct answer.**

Hydrostatic pressure and shut in pressure will affect the bottom hole pressure.

BHP = Hydrostatic Pressure + Shut in Pressure

Go to next question.

Question - 9: The well is shut in with 450 psi.

Current mud weight = 11.5 ppg

Well depth = 8,000'MD/7,500'TVD

What is the formation pressure, psi?

a) 4,532 psi

b) 4,935 psi

c) 5,234 psi

○ **This answer is incorrect.**
Please try again.

Go back to the question#9

This is the correct answer.

Formation pressure = Hydrostatic Pressure + Shut in Pressure

Formation pressure = (0.052x11.5x7500) + 450 =4,935 psi

Go to next question.

Question - 10: The well is shut in with 450 psi.

Current mud weight = 11.5 ppg

Well depth = 8,000'MD/7,500'TVD

What is the equivalent mud weight of formation pressure?

a) 12.58 ppg

b) 12.65 ppg

c) 12.85 ppg

This answer is incorrect.

Please try again.

Go back to the question#10

This is the correct answer.

EMW = Current Mud Weight + (Shut in casing pressure $\div$ 0.052 $\div$ TVD)

EMW = 11.5 + (450 $\div$ 0.052 $\div$ 7,500) = 12.65 ppg

<u>Go to next question.</u>

Question - 11: Formation pressure = 4,200 psi.

Formation depth = 6,000'MD/5,500' TVD

Drill the well with 13.0 ppg.

What is the shut in casing pressure at this depth?

a) 482 psi

b) 354 psi

c) 144 psi

This answer is incorrect.

Please try again.

Go back to the question#11

 **This is the correct answer.**

Hydrostatic pressure = 0.052x13x5,500 = 3718 psi.
Shut in casing pressure = Formation Pressure - Hydrostatic Pressure
Shut in casing pressure = 4,200 - 3,718 = 482 psi.

Go to next question.

Question - 12: 15 bbl of gas bubble migrates up 2,500' TVD in 9,000' TVD well while the well is shut in. Hole section is 12-1/4" and mud weight is 10.5 ppg. What is the approximate increase in casing pressure?

a) 1,250 psi
b) 1,300 psi
c) 1,365 psi

This answer is incorrect.

Please try again.

Go back to the question#12

This is the correct answer.

Increase in casing pressure = 0.052 x 2,500 x 10.5 = 1,365 psi

Go to next question.

Question - 13: Pump pressure at 35 spm is 1,500 psi. What is the pump pressure at 30 spm?

a) 1,102 psi

b) 1286 psi

c) 1421 psi

○ **This answer is incorrect.**

Please try again.

Go back to the question#13

 **This is the correct answer.**

New pump pressure = Old pump pressure x (New SPM ÷ Old SPM)^2

New pump pressure = 1500 x (30 ÷ 35)^2 = 1,102 psi

Go to next question.

Question - 14: Pump pressure at 25 spm is 1,000 psi. What is the pump pressure at 50 spm?

a) 2,000 psi
b) 3,000 psi
c) 4,000 psi

This answer is incorrect.

Please try again.

Go back to the question#14

This is the correct answer.

New pump pressure = Old pump pressure x (New SPM ÷ Old SPM)^2

New pump pressure = 1000 x (50 ÷ 25)^2 = 4,000 psi

<u>Go to next question.</u>

Question - 15: With 13.0 ppg mud weight, pumping pressure is 3,000 psi. What is the pumping pressure with 14.5 ppg mud?

a) 3,124 psi

b) 3,346 psi

c) 3,732 psi

This answer is incorrect.

Please try again.

Go back to the question#15

This is the correct answer.

New pump pressure = Old pump pressure x (New MW ÷ Old MW)

New pump pressure = 3,000 x (14.5 ÷ 13.0) = 3,346 psi

<u>Go to next question.</u>

Question -16: If a 15 bbl of gas bubble at 4,000 psi is allowed to expand to 30 bbl, what will be the bubble pressure?

Note: Neglect temperature

a) 3,500 psi

b) 3,000 psi

c) 2,000 psi

This answer is incorrect.

Please try again.

Go back to the question#16

This is the correct answer.

Use Boyle's Gas Law: P1 x V1 = P2 x V2

P2 = (4000 x 15) ÷30 = 2,000 psi

Go to next question.

Question -17: A 15 bbl kick in larger hole creates less Shut In Casing Pressure (SICP) than a 15 bbl kick in smaller hole. What is the reason?

a) Height of influx is less in a large hole.

b) Bigger hole creates less annular pressure loss.

c) Large hole has less restriction than smaller hole.

This answer is incorrect.

Please try again.

Go back to the question#17

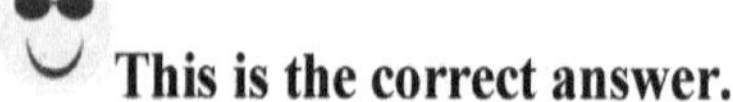 **This is the correct answer.**

Height of influx in the annulus is less in a larger hole therefore there is less reduction in hydrostatic pressure which result in lower SICP.

Go to next question.

Question -18: Gas cut mud reduces mud density from 13.0 ppg to 12.8 ppg.

Well depth = 12,000'MD/10,000' TVD

What is the reduction of hydrostatic pressure at the bottom hole?

a) 104 psi

b) 125 psi

c) 130 psi

This answer is incorrect.

Please try again.

Go back to the question#18

 **This is the correct answer.**

Reduction in hydrostatic pressure = (13.0-12.8) x 0.052 x 10,000 = 104 psi

Go to next question.

Question -19: The well is planned to TD at 9,500'MD/9,000'TVD.
Expected formation pressure at TD is 4,500 psi.
Plan to have over balance 300 psi over formation pressure.
How much mud weight should be?

a) 9.7 ppg

b) 10.3 ppg

c) 11.0 ppg

This answer is incorrect.
Please try again.

Go back to the question#19

This is the correct answer.

Desired hydrostatic pressure at the bottom = 4500 +300 = 4800 psi

EMW = 4,800÷0.052÷9,000 = 10.3 ppg

Go to next question.

Question -20: Which two of the following do not increase with gas migration in shut in well?

a) Gas bubble pressure and pit volume

b) Bottom hole pressure and pit volume

c) Casing pressure and drill pipe pressure

This answer is incorrect.

Please try again.

Go back to the question#20

This is the correct answer.

Gas bubble pressure and pit volume will not increase in shut in well.

Go to next question.

Question -21: The well is shut in with 6-3/4" drill collar (3.34" ID) and shut in casing pressure is 350 psi. What is the force pushing the drill collar upward?

a) 12,525 lb

b) 9,568 lb

c) 1,752 lb

This answer is incorrect.

Please try again.

Go back to the question#21

This is the correct answer.

Force = Pressure x Area

Force = $350 \times 0.7845 \times 6.75^2 = 12{,}525$ lb

Go to next question.

Question -22: The well is shut in and drill pipe pressure is 350 psi. Current mud weight in hole is 9.2 ppg and well depth is 8200'MD/7,500'TVD.

What is the kill weight mud?

a) 9.8 ppg

b) 10.1 ppg

c) 11.1 ppg

This answer is incorrect.

Please try again.

Go back to the question#22

This is the correct answer.

KWM = Current Mud Weight + (SIDPP ÷0.052÷TVD)

KMW = 9.2 + (350 ÷0.052÷ 7,500) = 11.1 ppg

Go to next question.

Question -23: The plan is to cement 7"casing in 9.5" hole and 35 bbl of sea water is pumped ahead as spacer.

How much bottom hole pressure will be reduced when all of sea water is out of the shoe?

Shoe depth = 8,000'MD/8,000'TVD
Mud weight = 10.0 ppg
Cement weight = 14.0 ppg
Water weight = 8.6 ppg

a) 45 psi

b) 64 psi

c) 83 psi

This answer is incorrect.

Please try again.

Go back to the question#23

 **This is the correct answer.**

Annular capacity = $(9.5^2 - 7^2) \div 1029.4 = 0.04007$ bbl/ft
Length of seawater in the annulus = $35 \div 0.04007 = 874$ ft
Hydrostatic pressure reduction = $(10.0 - 8.6) \times 0.052 \times 874 = 64$ psi

<u>**Go to next question.**</u>

Question -24: Fracture gradient at shoe = 0.65 psi/ft. Casing shoe depth is 8,500'MD/8,500' TVD.
Current mud weight is 10.0 ppg.
Well depth is 12,000'MD/12,000'TVD.
What is the maximum annular pressure loss before fracturing formation at shoe?

a) 1,105 psi

b) 1,200 psi

c) 1,480 psi

This answer is incorrect.

Please try again.

Go back to the question#24

 **This is the correct answer.**

Fracture gradient in ppg = 0.65 ÷ 0.052 = 12.5 ppg

Annular pressure loss = (Fracture Pressure - Mud Weight) x 0.052 x TVD

Annular pressure loss = (12.5-10.0) x 0.052 x 8,500 = 1,105 psi

<u>Go to next question.</u>

Question -25: Current depth = 11,000' MD/10,000' TVD.
Formation pressure gradient = 0.52 psi/ft
What is surface pressure if the well is full of gas (0.1 psi/ft)?

a) 4,200 psi

b) 4,620 psi

c) 4,830 psi

This answer is incorrect.

Please try again.

Go back to the question#25